Aquaponics System

A Practical Guide To Building And Maintaining Your Own Backyard Aquaponics

Bowe Packer

Bowe Packer

TABLE OF CONTENTS

Publishers Notes	4
Dedication	6
Getting The Most Out Of This Book	7
Part I: Introduction	9
Chapter 1- Why Give it a Try?	11
Chapter 2- Some Things to Know about Aquaponics	15
Chapter 3- Solar Ponds	18
Chapter 4- A Variety of Choices	20
Chapter 5- Playing with the Water	23
Chapter 6- What to Know about Plastic Tanks	26
Chapter 7- Keeping all the Pieces	28
Part 2: Setting up at Home	32
Step 1 – Cutting the Opening of the Holding Tank	33
step 2 – Creating Barrel Stability	34
Step 3 – Cutting the Barrels	37
Step 4 – Creating the Cuts for Pipework	39

Step 5 – Final Cutting for a Duckweed Tank   40

Step 6 – The Piping System   43

Chapter 8- Operating your System   45

Chapter 9- Frequently Asked Questions   48

Chapter 10- Aquaponics Produce   56

Chapter 11- Native Aquatics and Foreign Aquatics   57

Chapter 12- Staying in Balance   59

Chapter 13- How Many Fish Can it Hold?   60

Chapter 14- Why to Keep Your Fish Low   62

Chapter 15- Keeping Everyone Healthy   64

Chapter 16- Growing Your Grow Beds   66

Chapter 17- Troubleshooting Your System   69

About The Author   78

Bowe Packer

PUBLISHERS NOTES

Disclaimer

This publication is intended to provide helpful and informative material. It is not intended to diagnose, treat, cure, or prevent any health problem or condition, nor is intended to replace the advice of a physician.

Also, please understand that this guide is intended to help get you off to a great start of learning about aquaponics systems. You will run into things I did not, that is just the natural process of life.

No action should be taken solely on the contents of this book. Always consult your physician or qualified health-care professional on any matters regarding your health and before adopting any suggestions in this book or drawing inferences from it.

The author and publisher specifically disclaim all responsibility for any liability, loss or risk, personal or otherwise, which is incurred as a consequence, directly or indirectly, from the use or application of any contents of this book.

Any and all product names referenced within this book are the trademarks of their respective owners. None of these owners have sponsored, authorized, endorsed, or approved this book.

Always read all information provided by the manufacturers' product labels before using their products. The author and publisher are not responsible for claims made by manufacturers.

Paperback Edition 2013

Manufactured in the United States of America

Bowe Packer

DEDICATION

I dedicate this book to all those people out there who remind us of the things we have forgotten about ourselves.

And this holds especially true of my beautiful and amazing wife, Alma. She is the one woman who has the most amazing talent to let me grow and love the things about myself that I have not fully accepted.

I cherish the love she has for me when I may not know how to love myself.

May we all have this kind of beautiful soul in our life.

Sent from LOVE,

Sunshine In My Soul

GETTING THE MOST OUT OF THIS BOOK

If you, like many others, really want to create an aquaponic system that you can use in your own home then this is the perfect book for you. We're going to teach you, right here, in two very simple sections, everything that you need to know to get started on your own project. It's much easier than you may have thought and by the time you're done here you'll know more than you ever thought you'd need to know (but it's all important).

The first section of this book discusses why you should be trying out aquaponics for yourself. Of course if you've picked up this book then you're already interested but *why* are you interested? Do you know all the cool things that you can do in your own home or all the reasons to get started? Well this section will tell you about exactly those things and it will also tell you how aquaponics works and why it's different from hydroponics.

Throughout the various chapters of this book you'll also learn about the way an aquaponic system works and some of the different types that you could explore for your own uses. You'll even learn about water flow systems, the types of containers to use when creating your system and finally how to build it.

Bowe Packer

Section two is all about the building process. This is where you'll find out how to setup all the pieces perfect for wherever you live. If you have any questions you can even look them up in our FAQ's area where you'll likely be able to find the answer. If you don't then check out the troubleshooting section next for the answers to all of your aquaponics questions.

PART I: INTRODUCTION

For myself the journey into aquaponics started not that long ago. In fact it was only a few years ago that one of my close friends first mentioned the word 'aquaponics' to me. I was skeptical of course much like your friends and family may be but then I decided that I could give it a try. How hard could it be raising fish, vegetables and fruits after all? Well I was in for a bit of a surprise I will definitely admit.

I was intrigued by the idea just like you probably are. I couldn't imagine the level of self-sufficiency that it would take for an aquaponics system to work. And thinking of all the great products I was going to get from fish to various types of fruits and veggies made the work seem like it was definitely going to be worth it. This was going to be easier than having my own farm or garden yet it would give me even more results. So I set out to give it a try.

I got my first blueprint from the same friend who mentioned aquaponics to me in the first place. He actually realized that he wasn't going to be able to do it himself but figured I might still be interested. So I took the blueprint and started on my way. It may seem strange but I was actually off and running faster than you may have thought possible (and definitely faster than

Bowe Packer

I thought possible). This is where I'm going to start sharing my journey, and all the things I've learned, with you.

Chapter 1 - Why Give it a Try?

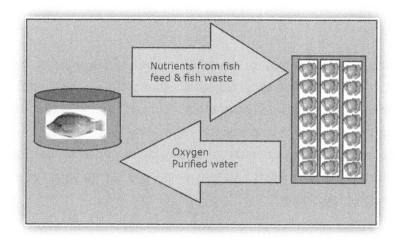

Okay so the first thing to think about is why you should try it out at all. After all this is going to take a lot of hard work and effort on your part so you want to make sure there's a good reason for trying it out and there is. You will have a completely sustainable source of food that also cleans itself. How's that for a good reason?

For most people it's nearly impossible or at least difficult to have a farm because your home simply isn't big enough. You may have thought about farming before but your yard wouldn't accommodate enough that you felt it would be worth it. So you gave it up and continued shopping at the grocery store instead of growing your own food even though it's cheaper and healthier. With an aquaponics system you don't need to worry about the space because it's actually relatively small.

Aquaponics is a type of aquaculture and hydroponics system that's sort of mashed all together. Aquaculture is farming fish and hydroponics is vegetable farming with water. So aquaponics is fish and vegetable farming with water. But of course that doesn't make a whole lot of sense all by itself either. So here's what happens when you set up your system and get it working.

1. Freshwater fish are put in holding tanks with aerations systems so that the water is continuously oxygenated.
2. You can feed the fish with any type of pellet or organic food.
3. When the fish produces waste it becomes mixed with the water in the tank as well as unused fish food.
4. The water from the fish tank is pumped into growing beds where your vegetables are planted.
5. The nutrients in the water that is pumped into the growing beds helps them to grow and the soil purifies the water.
6. The purified water that isn't used in the beds is pumped back into the holding tank where the fish are located.
7. The cycle continues for as long as you keep planting vegetables and feeding your fish. So you can have fresh

products all the time. It's a great way to make sure you're eating healthy all the time.

So what does your aquaponics system have that helps it to work so well? It has three things that all work together.

- **Edible fish**
- **Bacteria (in the water from the fish)**
- **Plants (vegetables)**

Bacteria grow just about anywhere but when you start feeding your fish they have to get rid of waste somewhere and that just so happens to be inside their tank. When the waste is expelled different types of bacteria begin to grow. One of those works to breakdown the waste materials and creates nutrients that are used by the plants in your growing beds.

So how is an aquaponic system different form an aquaculture system? Well you've probably heard of aquaculture before but here's a quick rundown of how it works. These systems use freshwater fish in a big two thousand (or so) liter tank. If you keep a bunch of fish in that large tank you have to get rid of approximately 200 liters of the water each day (and then replace it) because of the waste involved. An aquaponic system uses the waste as a fertilizer.

With that same two thousand liter tank you can have four growing beds at one time. That tank and those beds will produce approximately 70 kilograms of vegetables (or more) and 40 kilograms of fish (or more) in only six months. It would take a lot longer for your aquaculture system to produce that much and you'd have to be pumping out and replacing a lot of water while you're at it. So aquaculture would take a lot longer too.

Chapter 2 - Some Things to Know about Aquaponics

1. In only six months your fish could grow from about 50 grams to a whopping 500 grams if you're feeding it well. Of course this will also depend on the specific type of fish that you choose to raise as not all fish grow this fast or this large.

2. Supporting your family with an aquaponics tank and the fish/vegetables that it produces means you're going to need a pretty good size tank. Approximately 3000 liters is probably ideal though you may be able to get by with 2000.

3. A holding tank that's approximately the size we're talking about (2000-3000 liters) will hold about 100 freshwater fish. They don't all have to be the same type of fish but make sure they're all conducive to the right type of water.

4. Make sure that you don't put your growing beds on the ground. There are many types of insects and animals that will get into your vegetables if you leave them on the ground. Just think of the problems everyone has with regular gardens. By elevating yours you'll have fewer

problems. You could also choose to use pesticides though organic foods are even better to eat.

5. Solar panels can actually help you even more than using electricity because they save you money and store up that energy for you to use all the time. You won't have to worry about running out of energy and you won't have to worry about running up your energy bill. A 65 watt panel will help you with pretty much any tank and though the cost may seem daunting at first the amount it will save you is worth far more.

6. Now something else you should know is that an aquaponics tank does not have to be a custom made, special piece of machinery. Just about anything can be turned into a tank as long as it's big enough, sturdy, solid and free of cracks or holes. You can use pretty much anything you want so long as it's considered food-grade meaning it's safe for holding food (which is what you're going to be putting in it).

When you've got your entire system ready you simply need to talk a quick walk around each day to make sure everything is running smoothly and nothing needs replaced. That's all it takes is just a few minutes unless there's a major problem such as cleaning or replacing a part. Or when you're ready to harvest of course.

Chapter 3 - Solar Ponds

Now you may think that aquaponics is something entirely new but that's actually not the case. This is a type of growing that's been around for quite a while. After all, many people have been looking for less expensive ways to grow crops that they need for their families. The first way of doing this was through a smaller method (though similar) of using solar ponds.

A solar pond is much smaller than an aquaponics system. Typically this will be just one small fish tank. It's very similar in other respects but most definitely very small and the one tank supports both fish and vegetables. Generally the fish would be beneath a bed of lettuce with a mesh net between so the lettuce gets the nutrients from the fish but the fish don't eat the lettuce.

In this type of system the lettuce is put in the middle of the fish tank (usually a round tank). It then continues to grow as it receives more and more nutrients and gets moved to the outside where it will receive fewer nutrients. The smaller lettuce stay in the center. It's important to use a hardy fish in this type of tank.

Chapter 4 - A Variety of Choices

There are a few different methods of using aquaponics but here we will discuss the three most popular ones. You'll notice that they become increasingly complex and costly as you go down the list so think about what you're willing to do or not do for your system.

1. The most simple method of aquaponics is simply to dig a tank into the ground. All you do is dig out a hole and fill it with water and fish. You don't use this type to grow vegetables and essentially all you are doing is raising a cheap fish tank.

2. A solar pond like we discussed in the last chapter is the second method and is slightly more complex. This method uses a single fish tank which grows the fish and vegetables in the same tank.

3. Finally we come to the aquaponics system that we've discussed in earlier chapters. This is where one tank is dedicated to the fish and one is dedicated to the plants. This system also has a mechanism for transferring

water between each tank and for purifying the water as well.

Okay so now let's go back through and look at all of those options again. The first is simply a hole dug in the ground. Most of this could hold more than three thousand liters of water which means you're going to dig a pretty big hole to fill with fish and water. You may choose this method if you have a ranch or a farm already. Of course you can use it anywhere but these are the most common places.

There is no type of aeration or filtration in this type of tank which means only certain types of fish will be able to live there. You'll need to do a little research to figure out which fish those are because the only time your tank will be oxygenated is when it rains. The fish will be fed however as algae will grow over the water which they can eat. This is a good type of tank for those who simply want some fish to eat.

On to the second type of tank you'll be able to grow limited amounts of produce as well as the fish. This type of tank allows for the fish and vegetables to grow together which makes for a much smaller system. This method requires you to purchase or find a tub, barrel or other object which can hold a few thousand liters of water. You want to make it large enough to be worth your while after all.

Make sure that you are purchasing aquaponic plants so that they float on the top of your water and don't sink to the bottom. You won't get any produce that way. This raft method allows the plants to grow quickly and easily. Of course make sure you put a net underneath them so that your fish don't nibble on the roots and kill the plants before they can grow.

Finally we come to the aquaponics option that we've talked about throughout the first part of this book. This system uses two tanks; one for growing vegetables and one for your fish. It also has a filter which allows the water to be purified somewhat before going into the vegetable beds. This is a more complicated system (though not much) and most definitely a more controlled one.

CHAPTER 5- PLAYING WITH THE WATER

Now with an aquaponics system you want to make sure that everything is going to work smoothly and the best way to do that is to experiment a little. You could experiment with the types of fish you use and the plants that you choose to grow. You can also choose to experiment with your water flow.

Now your plants are going to need water to grow and your fish need new water periodically. So how do you take care of those needs? You're going to need to think about how often you want your plants to be watered so that you can set up your water pump to do what you want it to do.

Typically you'll have three options.

1. Hourly-With this method your water pump will remove whatever amount of water you tell it to every hour. This water comes out of the holding tank and feeds into your grow beds. The water from the grow beds will seep through the ground and through drain holes that never close and from there back into the holding tank.

2. Continuous-With this method your water pump is always working. Instead of taking a larger amount of water out of your tank at one time it takes out small amounts continually and passes them on to the grow beds. The grow beds still have drain holes that allow the water to pass back through into the holding tank.

It may seem like a good idea to use continuous water flow for your grow beds but that is actually not the case. If you've ever had a garden you know that plants don't receive a continuous supply of water and they can drown if you give it to them. So if you're going to use a continuous supply method you need to have plants that are hardy and able to withstand the high amounts of water.

3. Draining-The final method is called draining and it uses water added into the holding tank. The holding tank is filled over with water which floods the growing beds. When the growing bed receives the large amount of water all at once it slowly filters that water through and back into the holding tank. This method, similar to continuous watering, needs to be monitored so that the plants are not flooded too often where they die out.

Chapter 6 - What to Know about Plastic Tanks

When you're growing anything in plastic you want to make sure that you're using what's called food grade plastic. There's a very important reasons for this in that you want to eat the foods that you're growing. With regular plastic (not food grade) there are harmful chemicals used in the production process. Those chemicals can leak into your food and make you and your family sick.

Another thing people wonder about is metal tanks. It's definitely not recommended that you use any type of metal to store your fish or vegetables in either as this can cause rust in the water. It can also harm your fish as the chemicals from the metal get into the water itself and thereby into the fish. You may choose to use stainless steel if you're absolutely set on a metal system.

Finally there's recycled content. Now recycled products are great and it's wonderful to recycle anything you can. The problem with using recycled plastic for your tank is that you may not know 100% where it came from or what it was used for. Try to use only plastics you know used for neutral contents, something that won't hurt you or your fish. This is because those products can become very dangerous when the fish ingest them, which they will as the chemicals get released into the water.

If you must use something that you are unsure of then make sure you coat it with a liner so that any chemicals that may be present are kept out of your water and away from your fish. Also try to coat the inside and outside of your tank so that the environment does not cause too much wear and tear too quickly or you'll have to replace your tank a lot sooner than you want to.

Chapter 7 - Keeping all the Pieces

So what all are you going to need with your aquaponics system? There are actually quite a few components that you *could* choose though you are not required to use many of them. In fact, your tank will work just fine with only a few bare essentials but if you're looking to get more done for less work then you may want to consider a few additional pieces.

1. Automatic Feeder-Now you can, of course, choose to feed your fish every day (a few times a day) by hand. You don't have to purchase a feeder mechanism. This is, however, a great tool for someone who doesn't really want to sit around their house all the time. You'll be able to go out whenever you want and not have to worry about your fish being fed.

2. Automatic Siphoning System-A siphoning system is responsible for keeping your tank at the appropriate water level. This system will help to get rid of the water when the level gets too high. The problem with this is that you can't simply purchase them anywhere. You will need to build your own siphoning system if you choose to use one.

3. Bacteria- The third thing which is absolutely necessary in your aquaponics tank is bacteria. Without bacteria

Aquaponics System

your vegetables and fish won't thrive as well. Of course when you run your system properly bacteria is naturally produced which means you don't need to worry about adding this. However if you choose to add more make sure you use water from a river or pond nearby and simply add it to the water in your tank.

4. Make sure you don't put chlorinated water into your tank or add any chlorine as this will kill of the bacteria and your fish. It will also put chemicals into your vegetables making them less safe for you to eat.

5. Filter Media-Next you may want to think about a filter media for your tank. A filter media is something that keeps large pieces of bacteria or other products from going into your plants. This could be anything from regular gravel or pebbles to limestone. Of course the use of limestone can be bad for the ph level of your water because of the acidity so be careful with this one.

6. Batteries-If any part of your system uses electricity make sure that you are prepared in case you lose power. You want to have a system that is capable of running no matter what happens or you could be in trouble when the power goes out and there's no water pump working in your tank.

7. Fish-Of course in order for you to be able to do anything with your tank it's going to need to have fish

in it. You want to make sure that you're doing your homework and selecting the best possible fish for your needs and the conditions of your environment. If you don't have a lot of space then don't get fish that need a lot of space or grow too large. Don't get warm weather fish if you live in a cold climate.

8. If you've never raised fish then you want to look for ones that are going to be easier to raise. You want something that doesn't take a whole lot of care to start with so you don't lose out big when you make a mistake. Remember this ration, ¼ pound of fish for 1 gallon of water. Try not to exceed this ratio or you could have problems with your tank and your fish.

9. Remember that too many fish in a small area will cause problems with their growth and may cause them to kill each other off. You don't want to pay a lot of money for a bunch of fish only to lose most of them because your tank is too small. Think about the size you have available before you start buying anything.

10. Food-You obviously can't raise a bunch of fish without buying some fish food. So make sure that you are getting the best you possibly can. Remember also that the food your fish eat is going to partially make its way to the vegetables that you're growing as well so you'll want as much organic products as possible.

11. Feeder fish, crickets, mealworms and organic fish food are your best options for the fish and for your

vegetables as well. If you don't have the time to grow or collect natural types of food for your fish they can actually eat dog food as well. Most dog food contains a lot of vitamins and minerals that are helpful to the fish and your plants.

12. Tank-Finally you need to have your holding tank. You want to make sure that it's made of food-grade plastic or at the very least stainless steel. You also want to make sure that the material is heavy enough to hold several hundred liters of water for a long period of time and that it has no cracks.

Make sure that you use a good liner and protectants on both the inside and outside of your tank especially if you're using one that's origin is questionable. You don't want any chemicals getting into your water. You also don't want your tank to erode due to the weather conditions.

Part 2: Setting up at Home

So now you're ready to set up your own tank. You've read about what an aquaponics tank is and you know about the different versions. You also know what things you're going to look out for. So let's get started setting up your system. We're going to show you how to set up a very simple aquaponics system using two tanks. You may need to adjust some of it as you work to make it work best for you but for now just follow along. You're going to need a few different materials to get started:

- 3 Food-grade barrels (20-55 gallon)
- 8 2 x 4
- 6 2 x 10's
- 6 1 x 8's
- 1 8-foot hose
- 1 water pump capable of pumping 800 gallons per hour
- 1 10 foot long 1in. PVC pipe
- 7 "L" pieces pipe
- 1 "T" piece pipe

Aquaponics System

STEP 1 – CUTTING THE OPENING OF THE HOLDING TANK

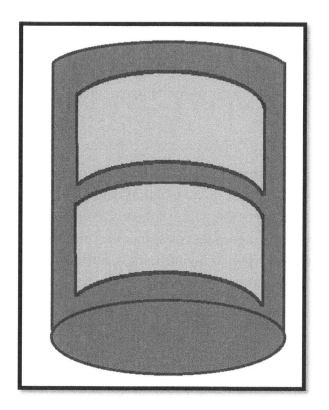

So the first thing you need to do is cut the opening of what's going to be your fish tank. You can choose how large you want it to be though it's generally a good idea to make it as large as possible so you can see what's going on in the tank easily and you can reach in if you need to grab something. Of course that doesn't mean you should cut out the entire center of your barrel.

The strip that you can see right down the middle of this design is actually important. Remember you're using a plastic barrel to create your tank. This means that a lot of pressure will cause the tank to warp causing you problems further down the line. That strip of plastic in the middle helps to keep the sides of the tank from falling in.

Another great thing about that strip is you can use it to install a screen later if you feel that neighbourhood creatures may get into your tank. A screen keeps animals from getting in, keeps leaves and debris out and doesn't impinge the oxygen that your tank is getting and the natural evaporation and rain.

STEP 2 – CREATING BARREL STABILITY

Now the next thing you need to do is make sure that your barrel is going to stand. Of course you cut a giant hole in one side so that means you need it to rest steadily on a rounded side. That's definitely going to take a little bit of work. What you're going to need to do is create a stand that will hold the barrel straight. This means the barrel needs to not rest on top of the pieces but actually sit down in them.

What you will want to do first is trace out the curve of the barrel onto your 2 X 10's before you cut them out. This way you'll have a perfectly rounded shape. You'll want 4 pieces of 2 x 10 so you can have two on each barrel (remember you have two barrels). You also want to make sure the pieces still have at least 2 inches below the arc of the barrel. Next you'll take the 2 x 4's and use those to secure the base.

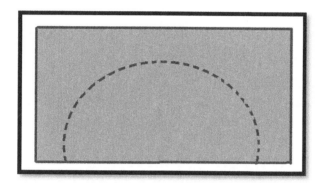

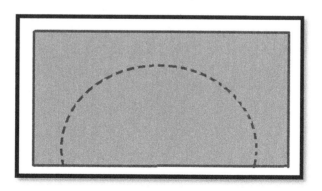

Using the barrel as a template cut a semi circle on the 2x10's.

Bowe Packer

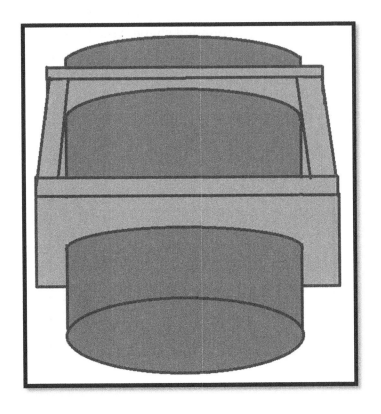

Add the 2 x 10s to the barrel and lock down the base of the barrel with some 2 x 4s. Voila!

Aquaponics System

STEP 3 – CUTTING THE BARRELS

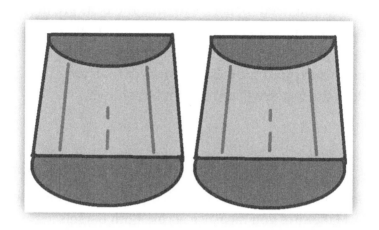

Cut the barrels lengthwise, as shown here

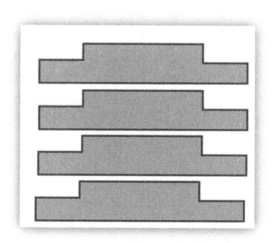

Create rectangular notches on the 2 x 4s

Bowe Packer

Now your holding tank is complete and ready to be set up. Of course first you'll need to get your grow beds ready for planting. You need to make sure that the beds are the right height for your holding tank so make sure that you measure your holding tank all the way to the top. You want to make sure your grow beds are taller than your holding tank so the water will drain right back down.

The grow bed is made from cutting one of your barrels lengthwise. This means you'll have two half barrels side by side. Make sure you measure across both barrels diagonally so you know how big to make your stand. You don't want it too small to hold your barrels or too big either.

Add four inches to the size you come up with for the width. Now you're going to cut rectangular notches out of the remaining 2 x 4's and then cut the 1 x 10's to create a platform. You'll also want to measure the length of your barrels and create supports out of plywood or leftover 1 x 10's so your platform stands sturdy.

Step 4 – Creating the Cuts for Pipework

So now your tank is completely stable and your growing beds are ready. What you need to do is create the pipework that's going to make your barrels into something that will actually transfer water and do the job it's supposed to do. But first you want to drill holes into your grow beds.

You want approximately 1.25 inch holes in the bed but remember that your pipe needs to fit in tightly. You don't want holes that are too big because the pipe will move around and your water will leak out. Put the female slip fittings into the holes in the grow bed and seal them tightly so there won't be any leaks. It will be a tight fit to get the fitting in but that doesn't mean you should skimp on your sealant.

Step 5 – Final Cutting for a Duckweed Tank

Now we come to step five. This is an optional step though it is recommended for those who are really serious about making their tank work. A duckweed tank will help oxygenate the water in your holding tank and it will also feed your fish so you don't have to purchase additional food. These are the best reasons for adding one into your assembly though you can skip over this section if you prefer not using a duckweed tank.

Make sure your platform for your grow bed is measured accurately so that you can make your duckweed tank the right height. You want to cut 1/3 of the barrel out by cutting crosswise. Seal up the ends of the barrel so that they are sturdy and they won't come off. You'll also want to add some sealant so they don't leak.

Next cut the rest of the barrel so that it has three support points like we showed above. You want the smaller part of the barrel on top of the larger and then nail them together. Make sure that all the holes or attached portions are sealed together with a heavy duty silicone sealant. Use the same plans for a platform that we gave in step 4 but make sure you measure the new dimensions properly.

Aquaponics System

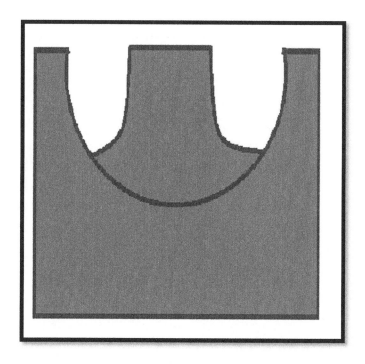

Cut the remaining 2/3 of the barrel so that three support points will remain

Bowe Packer

This is what the duckweed and dump tanks look like after both halves have been joined

STEP 6 – THE PIPING SYSTEM

The pipe system for your aquaponics system will require a bit more work and it's actually recommended that you hire a plumber to come in and help. This is because your pipes need to be installed properly or your entire system simply won't work and that's definitely not something you want to deal with.

Here's a general idea of what you want your end product to do so you can tell the plumber and they'll be able to make sure it works the way you need it to.

- The water is going to come out of the holding tank and into the other tanks. You need the water pump therefore to be connected to the holding tank and not one of the others.

- You want to make sure your siphon system takes water from the holding tank to the duckweed tank then the dump tank and finally to the grows beds. Any excess water should flow into the dump tank.

- You need two pipes to lead from the holding tank into the grow beds.

Bowe Packer

- You need a separate pipe to return water from the grow beds if there is too much. That water should go to the holding tank.

You want gravel poured in once the main lines are installed.

Chapter 8- Operating your System

Make sure the water you put into your tank is non-chlorinated because too much chlorine makes it difficult or impossible for bacteria to grow. This means you want to get the proper water and not just water straight out of your sink. That water is actually full of chemicals that are not helpful to your fish either. Using water from a pond or river can help you to get started building the bacteria you need.

You want to be sure that the water you have cycles through all of your tanks every 40 minutes or so. This way you'll get all the water you need in each area and you also won't get too much. The first couple times this cycle occurs you want to make sure you're watching what happens. You need to know if there are any leaks in the system that need to be fixed and preferably before you add anything into the system. Just check the water.

Duckweed grows quite quickly and that's definitely helpful to you since it will feed your fish and allow the water to be purified without the use of products that will kill off your fish or cause problems in your vegetables. Make certain that you're checking the entire system frequently to find any problems before you get too far along on your system.

If you find something isn't working quite right for you with the system we've created then don't be afraid to make some changes. Everything we've created is open to adjustments and changes. You will be able to adjust whatever you have to make it work better for you.

Now you'll want to make sure you are raising the proper fish in your tank as well. There are two types of fish that you could raise depending on your needs and purposes. The first are ornamental fish. These are fish that you can't eat but they will help you produce great vegetables and they will look good swimming around your tank. The second are the edible fish that you can also cultivate and eat along with the vegetables.

Even with nonedible, ornamental fish you will be able to eat the vegetables that are produced. That's because any type of fish that you put into your tank will help to produce the waste materials necessary for your vegetables to grow better than ever.

So what types of fish can you put into your holding tank?

- **Tilapia**
- **Walleye**

- **Crappie**
- **Koi**
- **Pacu**
- **Carp**
- **Striped Bass**
- **Lake Perch**

Pretty much any freshwater fish will be able to live in your aquaponic system as long as you keep the water temperature appropriate. You're also going to be able to grow a wide variety of different plants in your grow beds from things like lettuce to tomatoes.

If your water gets to be polluted you will need to add new water so your fish don't get harmed by the pollutants. Make sure you're always watching out for these types of pollution to protect your fish and your vegetables from harm.

Chapter 9- Frequently Asked Questions

QUESTION: *Why is an aquaponic setup needed to cultivate plants without soil?*

So why can't you grow plants simply using a jar of water? Well the problem is that the plant needs a lot of nutrients in order to grow properly and those nutrient needs simply aren't met in a single jar.

QUESTION: *I like the home-sized system you provided, but is there any way to get the same results with the plants without having to raise the fish?*

If you want to grow plants without using the actual aquaponics system then you'll want to use something called manure tea in your water. This will help to produce the minerals that you need in the water so that your plants will grow well. This isn't considered aquaponics and it's actually strict hydroponics.

If you use a hydroponic system you need to make sure the water is continually drained so it doesn't get too much toxicity in the plants. This occurs as bacteria and chemicals build up in the water and your plants get hurt by those chemicals.

The aquaponics method of raising fish and vegetables uses the advantages of both traditional hydroponics and aquaculture to create a system that is better than either one by themselves. This is great for you and luckily it's also a relatively simple process to use.

QUESTION: *How come the roots of the plants don't rot even if there is direct contact with water?*

The roots of the plants are prevented from becoming too saturated by the filter medium that they are planted in which is typically rocks or gravel. They still stay moist continually but they are not sacrificed to a continuous puddle of water so the roots don't rot.

QUESTION: *I don't like the idea of pouring gravel into my grow beds. Is there any other way?*

A method known as D.W.C. or Deep Water Culturing enables you to grow your plants without the use of gravel. The plants are put into a piece of Styrofoam that floats on the water and the roots reach through into the water itself. Of course it's important to check the type of plants you're trying to grow to determine if they can survive constantly submerged in water.

QUESTION: *What is CHIFT? I've heard of this particular technique but I don't know what it is.*

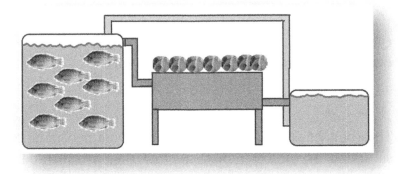

The CHIFT system. The dark gray pipes lead to the sump pit. The light gray pipe transport water back to the holding tank.

CHIFT means constant height of water in the holding or fish tank. This is one of the best systems that you can use because it's very simple to operate. You don't put the water pump into the holding tank and instead place it in the sump pit.

The illustration above shows that the holding tank should be higher than your grow beds. It also shows how the pipes should be working and where they should be situated. The sump pit collects water that isn't needed and then passes it back up to the holding tank. Your system is now capable of sustaining itself without needing any extra help from you.

So what are the reasons you should use this type of tank and system for your own aquaponics? Well there are actually two good reasons you should think about:

1. If the sump pump fails water will stay constant in the holding tanks even when none flows to the growing beds. The plants will survive a few more days and the fish will live for a while longer. In other systems the fish die immediately from the shrinking water level.

2. The system is easy to complete and can be done in only half a day.

There are two disadvantages to this system in that your holding tank must be large and elevated and your sump pit must be low yet still have a high capacity. This is because the sump pit must collect water that naturally drains from the grow beds rather than water that is pumped out. The holding tank must be the tallest and the sump pit needs to be the smallest.

QUESTION: *I like the idea of flooding a small grow bed with water from the holding tank. However, I want a system that does not require a sump pit. Is there a setup that is similar to a solar pond but still works like an ebb-and-flow aquaponic system?*

One way to do just this is to use a low holding tank with high capacity and an elevated grow bed for your produce. You need the grow bed to be situated above your holding tank so that when the pump activates the water from the holding tank goes straight into the grow bed and can then be drained back down. Because the plants are not submerged constantly there is less risk of rot. One problem is that the water in the holding tank lowers for a period when the grow beds are flooded.

QUESTION: *What are the advantages of keeping specific species of fish when I decide to set up an aquaponic system at home?*

You want to make sure that the fish you choose are best situated to the type of climate that you live in. A warm climate large fish is a barramundi. These are edible and they grow relatively quickly. They also tend to do well in warmer climates rather than colder ones so try to avoid these if you live in a cold area.

A catfish grows quite fast so they are great for a lot of different aquaponics systems. As long as you keep your water clean you won't have any problems. Dirty water however is not conducive to growing these fish. Though they can survive the dirt and pollution catfish definitely will not thrive that way. Because these fish have no scales they are also easier to prepare when it's time to harvest.

Carp are generally not accepted in the United States because they are considered an invasive species so it's important to check with an agriculture office in your area to find out if they are available and if you are allowed to cultivate them in your own aquaponics system. You may find that this is not allowed.

If you want to breed fish in your system then the best option is the simple goldfish. These are great for breeding once you add a layer of plants over the top of the water. This could be duckweed and once it's in place your goldfish will grow quickly. Though they are edible most people use goldfish as an ornamental fish.

Another option is a jade perch which is edible and has a high level of omega 3 fatty acids which are good for the brain and heart. This type of fish can be great for you and your family and they also grow well in freshwater aquaponics holding tanks.

Koi are typically not used as edible fish and are sold as ornamental. Though they are a type of carp they generally aren't grouped that way. You can use these in your aquaponics tank and can even mix them in with other types of freshwater fish to increase the visual appeal of your tank if you so choose.

Murray cod are also a great fish especially because you can stock a lot of them in just one holding tank. They taste great, they grow quickly and you can have more than you would of most other fish. Of course you will want to make sure they are fed properly or they will actually devour each other so take care to have enough food. You'll also want to make sure you have enough time to cultivate your tank if you choose these types of fish as the holding tank and system may take a little adjusting.

If you are willing to wait a bit to get your fish then you could select silver perch. These taste great however they take at least 12 months before they get to a size where you would want to harvest and eat them. If you're willing to be patient then this may just be worth it for you.

Finally we come to tilapia. This type of fish is probably one of the most common for anyone with an aquaponics tank as they are good tasting and they are pretty good at tuning out chemicals and other things that could end up in your water.

These fish are very easy to raise and they'll eat just about anything that you put into the tank. Really they'll eat just about anything that they find in the tank as well. Not only this but it only takes about 4-6 months for these fish to grow to a full adult size and be ready for eating. The only important thing is that they need a constant water temperature and it needs to be warm.

Chapter 10 - Aquaponics Produce

You're getting a lot out of a bit of hard work at the beginning and some very simple work at the end. You won't need to do much once your system is set up and you'll have fresh fish and vegetables whenever you want them. Who doesn't think that's worth it?

Not only that but you'll get the great feeling that everything you're producing and the food your family is eating was produced and grown by you and your own hands. It's also healthier than the products that you're going to get from the store because it's toxin free and chemical free as well.

Chapter 11 - Native Aquatics and Foreign Aquatics

Something you will definitely want to know about is the different types of fish in your area. You can make sure that you are getting fish that are local and ones that are allowed by your local government. The best way to do this is to talk to a fish expert or someone who works with the department of natural resources.

Local species are going to be the best option for your aquaponics system because they are easier to find in your area and they are more likely to be allowed. You won't have to worry about getting in trouble either. This is because some species of fish are not allowed in your area or not allowed to be captured within your area. You'll also want to make sure that you're not releasing your fish into the wild even if you change your mind about your system.

If you decide you do not want your fish anymore you want to make sure that you are getting rid of them in an approved way. Flushing your fish does not necessarily kill them contrary to popular belief. Instead, this releasing fish into your local sewer system. Your sewer system is not closed either which means that those fish actually get into your local waterways including ponds or rivers.

Bowe Packer

Invasive species can get into water systems easily in this manner which is very problematic. You want to make sure this doesn't happen so if you change your mind about aquaponics make sure you discard all of your pieces and all of your fish in a proper manner. The best option is to sell them to someone who want to raise them or eat them. If this isn't possible contact an authority in your area and find out what they prefer for you to do with the fish.

Chapter 12 - Staying in Balance

So how do you raise your animals within your tank? That's right I said animals and not just fish. That's because you can actually raise mussels, prawns and other freshwater creatures within your freshwater tank. All of these creatures produce the wastes and other products that you need to keep your entire system working properly. They also thrive in the same types of environments as well.

You will probably want to get started with just fish and your vegetables so that you can understand what you're doing and learn what needs to be done before you start changing things. Then you can start making changes and adjusting the things you don't like or think could be better. Don't be afraid to make some changes. If your tank is large enough to support everything you have then there's no reason to worry.

Chapter 13- How Many Fish Can it Hold?

So how many fish should you keep in your tank? We talked about the best ratio for your tank but of course most people don't follow that. So what are the best possible ways that you can add fish into your tank? You can have a large amount of fish actually. The more fish that you add into the tank the more fish that will die before they get big enough to be worth eating. However, you can continue to add whatever you want.

Your holding tank has nowhere to go and the fish that are going into it have nowhere to go. This means that you need to keep the water from getting too polluted and you need to make sure that your fish are fed properly. If you add too many fish in you need to make sure that you're watching the tank and the way that it runs so you know if there is a problem.

Be prepared to make some drastic changes quickly if you notice a problem. You need to be sure that you are taking care of any problems that arise fast or you'll lose even more of your fish. If you notice that there are too many of your fish dying or your pollution level is going up too drastically make sure you start pulling fish out fast or you'll have even more problems as time continues to go on.

If you are prepared for problems you'll be fine if anything comes up later. More fish in your tank will mean a lot more problems. Don't forget to check that tank as frequently as possible so that you notice anything that happens as soon as it begins to happen.

CHAPTER 14- WHY TO KEEP YOUR FISH LOW

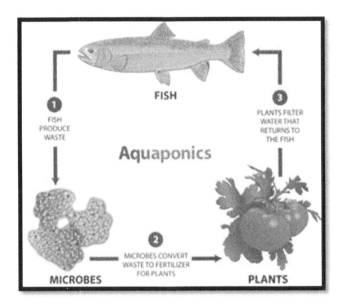

So why should you keep your fish tank low if there's not a lot of risk involved in having more fish? Well there are actually quite a few reasons.

- If you want to primarily harvest a lot of crops it's actually possible to grow a large amount without having to use a lot of fish. You'll be able to use only 9-10 fish to grow a medium growing bed.

- Less fish keeps the stress level of your fish down. When your fish are stressed they don't grow as well and they tend to die off faster. This can be a negative effect for anyone who wants to harvest their fish.

- Less fish means that you won't put too much food in the tank which means your fish don't eat too much.

- If your fish become ill it's better to have less fish because you have less loss. It's possible for your fish to contract illnesses from one another which means you lose even more money from the cost of those fish.

Fish in small systems grow faster than those who are in larger systems because this is the way they primarily grow and live in the wild and is therefore what they are used to and most comfortable with.

Chapter 15 - Keeping Everyone Healthy

It's important that your fish and your plants all stay healthy because that's where you're going to get all of your food. There are several different things you need to know about your system and there are also a few important concepts to remember.

- **Nutrient deficiencies happen in aquaponics**

- **Nutrient deficiencies need to be done in a way that aren't going to hurt your tank and the products that you are growing in the tank.**

There is no soil in your system when you use aquaponics which means that nutrient deficiency is actually very common. This happens very frequently when you don't feed your fish the right quality food. You want to make sure that you're getting the best quality food that you can so your fish continue to grow and prosper the best way possible.

If your system runs properly you'll be able to save money in the long run because you have healthy fish. If you're working with sick fish you could have a lot of trouble and you'll have a lot of cost involved in treating them. You'll have less cost in purchasing higher quality food then you will with replacing the ones that die from lack of healthy food.

If you change the way that your fish eat and they still have problems you'll want to add some minerals into the water. Make sure you talk to an agriculture expert so that you can find the best possible level of minerals and nutrients for your fish so they not only continue to survive but actually thrive and grow. Those agriculture experts will be able to help you get the proper levels and keep them healthy.

If you don't know what different nutrients or minerals are going to do to your fish or your vegetables then talk with an agriculture expert. They will know best what's going to help your fish and what's going to be healthy for all the components in your tank as well as you the end consumer. You don't want to add in a lot of deadly or harmful nutrients or bacteria to your tanks which will harm you.

Chapter 16- Growing Your Grow Beds

Now when you plant your vegetables you should know that you actually have a very good chance of getting amazing plants and great food that will support your family for a long time to come. But you can't select just any products to grow. You have to make the best choices for you and for your area. But of course you need to select what's best for this method of growing as well. There are several plants that grow best in this type of system including the ones listed below:

1. *Eggplants*
2. *Tomatoes*
3. *Chili plants*
4. *Bell peppers*
5. *Beans*
6. *Cabbages*
7. *Lettuce*
8. *Papaya*
9. *Peas*
10. *Beetroot*
11. *Carrots*

Of course you need to be sure that you're planting the right things. Not that you need to plant seeds. You can actually plant crops or seedlings as well as the seeds that you plant in a regular garden. If you plant seeds make sure that you're keeping them in the right climate and the right type of soil. You don't want to lose your crops before you even get started.

Your seedlings may seem like they are adapting and growing faster than the seeds that you plant but this is no reason to worry. This is simply the way that your products will grow and you don't need to bother about it. The seeds will start catching up and growing better than before.

If you're not interested in harvesting fish you may wonder if you should even bother using aquaponics and the answer is absolutely. You don't need to spend as much money on fertilizer or as much time cultivating if you use aquaponics rather than simple hydroponics. These systems actually grow up to four times faster than hydroponics plants simply because of the additional nutrients you'll receive.

Another great thing is that your plants get a consistent amount of nutrients at all times instead of you simply providing them with small amounts of nutrients through fertilizers. So what types of aquaponics plants can you use?

Curled lettuces	Sugar pods
Savoy spinaches	Onions
Grosse lisse tomatoes	Beefsteak tomatoes
Lebanese cucumbers	Fordhook beetroot
Cos lettuce	Rockmelons
Comfreys	Yarrows
Lemongrass	Sage
Coriander	Mizuna
Chives	Broccoli
Cabbages	Snow peas
Watercress	Celery

Chapter 17- Troubleshooting Your System

Legend: (P) = Problem; (A) = Answer

P: My plants are all wilting and dying. What's wrong with my system?

A: Wilting plants may be caused by extremely high or extremely low pH levels. Check the pH level of your water and check the tolerable pH level for the plant types in your grow beds.

If you have a continuous flow system in place, the roots of the plants may not be getting sufficient oxygen. Try switching to a once hourly system or ebb-and-flood system instead.

If this doesn't work, try cultivating another type of vegetable and see if the new seedlings survive the grow bed. Pests may also cause wilting so be on the lookout for large pests and small bugs that may not be immediately apparent to the naked eye.

When in doubt, contact an aquaponic specialist in your area and consult with him/her. If it's *not* wintertime and your plants are suddenly showing signs of imminent demise, the nutrient levels in your water may not be adequate.

You can immediately remedy this problem by adding more fish feed to your holding tank. However, increasing fish feed may cause cloudiness in your holding tank, so make sure that you cycle the water frequently. If all else fails, draining the holding tank partially and refilling it with clean, non-chlorinated water will help clear up the effluents.

P: It is wintertime here and my plants are all wilting.

A: If you have planted native vegetables and fruits in your grow beds, then the problem is probably plant nutrition. During winter, all animals (with the exception of the polar bear, perhaps) move more slowly because of the dropping environment temperature.

When fish move more slowly, they require less food. With less food, you have less ammonia in the water – and plants need the ammonia in order to survive.

To improve this situation, you can choose to add some heating to the holding tank to encourage the fish to become more active. When the fish become more active, consider adding more fish feed so the effluent level in the water will increase. If this fails, you have no choice but to reduce the number of plants in the grow beds.

P: The leaves of my plants are turning brown or yellow for no reason at all

A: Yellowing or browning leaves are characteristic signs of toxicity. In some cases, the nutrient-rich soup being pumped into the grow beds may have high levels of minerals – too high for plants to tolerate.

Plants are extremely sensitive to chemicals. Plant tissue succumbs almost immediately to high levels of mineral salts in the soil.

The same thing happens when there's a high level of mineral salts in the water. Inversely, yellowing or browning of leaves may also indicate *nutrient deficiency.* In such cases, add more fish feed to the holding tank to increase the nutrient levels in the water being cycled to the grow beds.

P: I see aphids eating my plants!

A: Sadly, this is Mother Nature's way of feeding aphids. Insects do not discriminate between wild-growing vegetables and cultivated vegetables so you have to take care of the aphid invasion immediately. There are two ways that you can deal with this problem without resorting to pesticides.

Your first option is to plant vegetables that will attract beneficial creatures such as lady bugs. The lady bugs will take care of the aphids for you. This is a good long-term solution but sadly, it may not work if you have an intense aphid invasion on your hands.

The second option is to purchase a pack of adult lady bugs from your local agricultural supply or plant nursery. Now, *be careful* when introducing lady bugs to your grow beds. Chances are, there are ants nearby.

Aphids and ants have a symbiotic (mutually beneficial) relationship and therefore, anything that threatens the aphids will be dealt with by the stronger and meaner ants. It will be best to introduce beneficial insects during the evening when insect activity is generally lower.

P: I see caterpillars munching on my vegetables!

A: Again, this is just nature's way of feeding the 'young ones'. You don't have to be mean to the caterpillars. If you can make a homemade garlic spray, use that.

If not, just put on your gloves and manually pick off the caterpillars. Don't squish the bugs though! Instead, put the caterpillars in a jar with some leaves so you'll have some live food for your fish. This way, nothing is wasted *and* you are able to turn the tables.

P: My plants seem to be stunted.

A: This is most likely nutrient deficiency. Check one of the solutions stated previously. If none of the solutions work, check the pH level of the water. If the water is too alkaline, that may be the cause of the slow growth of your aquaponic plants.

P: The water level in the holding tank is always below the optimum level.

A: It may be hotter this time of the year *or* your plants are growing at a faster rate. In either case, just add more water.

Make sure that the water has been de-chlorinated. You can also add a booster dose of pond water to ensure that the bacterial population in the water remains constant. Water loss of more than fifty percent in a single holding tank system may require cycling of bacteria.

P: The water in my system is extremely dirty and there are a lot of floating effluents in the water. I can barely see my fish!

A: This is a *bad* situation because your fish can die in a matter of days if the water has become so polluted. The most common cause of polluted water in an aquaponic setup is *too much* fish food being added to the water.

If this is not the case, then the buffer (media) and the filtration mechanism are not performing well. You can add an additional filtering mechanism to the *return line* to clean the water. Any cleared waste from the water *must* be removed manually. Do not return filtered waste to the system as this will pollute the water once again.

P: There is a thin layer of ice on top of the holding tank.

A: The weather has finally gotten the better of your system. Consider heating the holding tank and installing additional 'bubble makers' to create mini-currents in the holding tank. The extra water movement will help decrease the incidence of ice formation on the surface of the water. But before you do *any* of these steps, check if your fish are still alive!

P: There is foaming in the water, even if the water has been cycled adequately a few weeks before.

A: There should be no foaming in established aquaponic systems. The most common cause of foaming is household detergents.

Household detergents may have been introduced when you 'topped off' the holding tank with more water. To remedy this problem, drain fifty percent of the water content of the holding tank and add de-chlorinated water. Continue doing this every day until the foaming goes away.

P: There are dead fish floating in the holding tank.

A: Remove the dead fish quickly. An aquaponic system is not designed to deal with dead fish – the excess of ammonia will overwhelm the system and may also cause increased fish mortality.

P: All of the fish are *nearly* dead but are still gasping for air.

A: There's no point in prolonging the agony of the fish. Aquaponic farmers usually just flatten the heads of almost-dead fishes. This is really the only way to dispose of the fish.

If you let the fish die in the holding tank, the water will only become more polluted.

P: Fish in the holding tank are not acting normally; some are swimming sideways, some are not eating well.

A: If this is your first time to raise such fish species, you may have given the fish the wrong type of food. Consider switching brands.

P: Fish seem to be gulping at the surface of the water.

A: The oxygen level in the water is insufficient to support all the fish. Add additional aeration mechanism to keep the water well oxygenated. This should solve the problem.

P: Fish are jumping out to catch insects that are hovering near the top of the holding tank.

A: The easiest way to deal with feisty fish is to install a screen on top of the holding tank.

P: I see red worms in the tank!

A: Don't worry about the red worms. If you can collect the worms, do so. You now have free, live food to give to your fish.

Bowe Packer

About The Author

Hello, my name is Bowe Chaim Packer and I like to see myself as an open, *"**wear my heart out on my sleeve**"* kind of guy.

Some of the most important things to me in my life are:

- Laughing
- Kissing
- Holding hands
- Being playful
- Smiling
- Talking deeply with others
- Being loved
- Loving others
- Changing the world one person at a time (if my presence in your life doesn't make a difference then why am I here?) Hmmmmm, maybe that is a topic for another book. ;-)
- Learning from others (although often times I first resist). However, don't give up on me….
- Sharing ideas (no matter what they might be)
- Learning about others via most forms of contact.
- Traveling – hello, of course – almost forgot one of my favorite pass times.

Remember, LIFE is a journey for each and every one of us. We must never forget the things that are important to us or lose sight of what makes us happy.